Omar Elahcene
Mohamed Yacine Bendjedou

Description and operation of the foggara

Omar Elahcene
Mohamed Yacine Bendjedou

Description and operation of the foggara

ScienciaScripts

Imprint
Any brand names and product names mentioned in this book are subject to trademark, brand or patent protection and are trademarks or registered trademarks of their respective holders. The use of brand names, product names, common names, trade names, product descriptions etc. even without a particular marking in this work is in no way to be construed to mean that such names may be regarded as unrestricted in respect of trademark and brand protection legislation and could thus be used by anyone.

Cover image: www.ingimage.com

This book is a translation from the original published under ISBN 978-620-2-54401-6.

Publisher:
Sciencia Scripts
is a trademark of
International Book Market Service Ltd., member of OmniScriptum Publishing Group
17 Meldrum Street, Beau Bassin 71504, Mauritius
Printed at: see last page
ISBN: 978-620-2-93547-0

Description and operation of the fouggara

By

Dr. ELAHCENE Omar

&

Doctoral student
BENDJEDOU Mohamed Yacine

To my parents ;

To my wife ;

To my daughters Douaa, Lina, Maria and Imene;

To my in-laws ;

To all my family and my beautiful family.

Table of Contents

Introduction

In the Saharan environment, water is available, but in its great majority it is very weakly renewable. Water resources in the Saharan regions are most often poorly managed and require rational and integrated management with a view to sustainability. Their exploitation must obey rules specific to the Saharan regions.

Thanks to their ingenuity, technical skill and keen sense of observation, the oasis people were able to judiciously exploit the large water reservoirs and adapt them to the capture of water and its slow return to the surroundings of the Western Erg. The topography of the terrain and the hydrogeology were used to advantage without damaging the environment.

Thanks to the Foggara, the oases were able to fertilize an arid environment and contribute to the development of an agricultural ecosystem. In spite of a hostile environment, a hyper-arid climate characterised by frequent sand winds, low rainfall and stifling heat, man in the Sahara has been able to create an artificial environment for thousands of years: the Oasis.

This particular ecosystem in a desert environment, essentially composed of a water source, a palm grove and a ksar, has been transformed into a real desert palace, built out of earth, palm wood and rock. The oasis was able to seek and capture water by all possible means. Better still, he was able to create an ingenious system for collecting and distributing underground water with rudimentary manual means.

The hydraulic work, well known in the Sahara under the name of Foggara, consists of capturing water through a system of gently sloping galleries to drain it to the irrigation network of the palm grove.

1- Etymology "Foggara".

Some historians believe that the term Foggara comes from the Arabic word "El Fokr" (poverty), meaning that whoever digs a foggara finds himself obliged to invest so much in it that he ends up falling into need before benefiting from it. Still others think that the name foggara is related to the word "Fakra" (vertebra in Arabic), since the wells are aligned like a "Fakratte" backbone. The most correct designation seems to come from the Arabic word "Fadjjara" (to gush) which would indicate the outlet of water from the mouth of a canal.

2- History and origin of the foggara

The traditional groundwater catchment system "Foggara" can be found in several countries but under different names, it is called "Foggara" in Algeria, it is found in Pakistan and Afghanistan under the name "Kariz", In Iran "qanat", Yemen "Sahrij", Oman "Falej", Syria under the name "Kanawat", Morocco "Khettara", Tunisia "Ngoula" and Spain "Minas", similar additions can also be found in Azrabeidjan, Armenia and ancient Egypt.

The foggara is not a Touâtienne originality, history tells us that the Assyrians and Persians knew it for a long time and that the Romans used it in Syria. It is indeed in the kingdom of Ourartou around the lake of Armya that it appeared at the beginning of the 1st millennium. It was probably originally a simple dewatering technique, designed to evacuate the seepage water that threatened to flood any mining gallery penetrating an underground aquifer. The genius of the Ourartèen was to "divert" this banal mining technique to make it what it has become, i.e. a specifically agricultural technique for capturing a water table for irrigation. The indigenous pastoralist peoples who occupied the central Sahara were confined around natural water points, whose number and importance was decreasing. It was the invading tribes that brought about the perfected practice of groundwater use. It would be interesting to know how and by whom the foggara was introduced into the Sahara. In Le Touât in particular, there are local chronicles, which are sometimes questionable, and it would be prudent not to draw any definitive conclusions from them.

It seems difficult to pinpoint the exact period of the appearance of the Foggara. According to Goblot (UNDP, 1986), the Foggara originated in Iran. The qanat which supplied Ibril in Persia, was built at the end of the 7th century BC. which attests to its very distant origins. In the Algerian Sahara, the Foggaras would have been introduced in the 11th and 12th centuries by El Malik El Mansour who would have dug the first Foggara in Tamantit, a locality located 15 km from Adrar. The Foggaras were developed in the Touat-Gourara by Arab-Berber tribes on the basis of the slavery of local labour or labour from neighbouring regions (Mali, Niger, etc.) (Hassani, 1991).

3- Birth of a foggara

Several hypotheses concerning the point of impact of the birth of a foggara are proposed, and are cited :

3.1- A source

The foggara has its point of origin in a spring, which may have dried up or the co-owners wanted to increase the flow, so a trench was made in the aquifer. This trench was built in the direction of flow of the water table, starting from downstream to upstream. The progressive deepening of the trench, the difficulty of evacuating the cuttings and the possibility of building a tunnel in the sandstone layer led to the well, the well-digger then moved to the surface of the ground, he dug down to the water table and connected these different wells by an underground gallery.

3.2- Upstream wells

The owners who wanted to dig a foggara and after choosing a point upstream, dug the first well to ensure the availability of the resource, then the wells follow one another from upstream to downstream, connected by the gallery until the water emerges at the surface, the exit point is also chosen downstream by the co-owners.

3.3- Downstream wells

The co-owners dig the first well downstream, then a second well connected by a gallery, then a third well, and so on, the wells are dug multiple times as they sink into the aquifer. This theory is the most suitable, since in the first case, the origin of a spring is not admitted from a hydrogeological point of view, since the aquifer is composed for the most part of loose formations (sand, clay and sandstone), and in the second case it is very difficult to work from upstream to downstream, with powerful water inflows in the gallery dug upstream (Benhamza, 2013).

4- The elements of a foggara

The Foggara system is divided into two parts: collection and distribution :

4.1- Collection

Groundwater is captured by a gently sloping tunnel several kilometres long, which drains the water from the water table to the open surface. This gallery is equipped with several vertical shafts which are used for the maintenance and aeration of the Foggara.

4.1.1- The gallery

The gallery is the driving force behind the Foggara (Photo 01). It is composed of two parts. The first part is the seat of a flow under load, while the second part is the seat of a free surface flow. The length of the gallery can vary from 1Km to 15km, while the range of the flow is from 1l/s to 50l/s (Remini, 1999).

The Foggaras are aligned from East to West and are carefully laid out to avoid any possible drainage that could damage neighbouring old Foggaras. The distance between two galleries must be greater than 100 "kamas". The Kama corresponds to the length of two open and outstretched arms of a normal sized man, i.e. about 2 metres. The gallery is made up of several "n'fad", a term designating the tunnel between two shafts.

As a general rule, the average distance between two "n'fad" is around 13 metres. The gallery ends with the Aghissrou, which represents the part between the first shaft, counted from the exit, and the Majra. This part can be covered by flat stones.

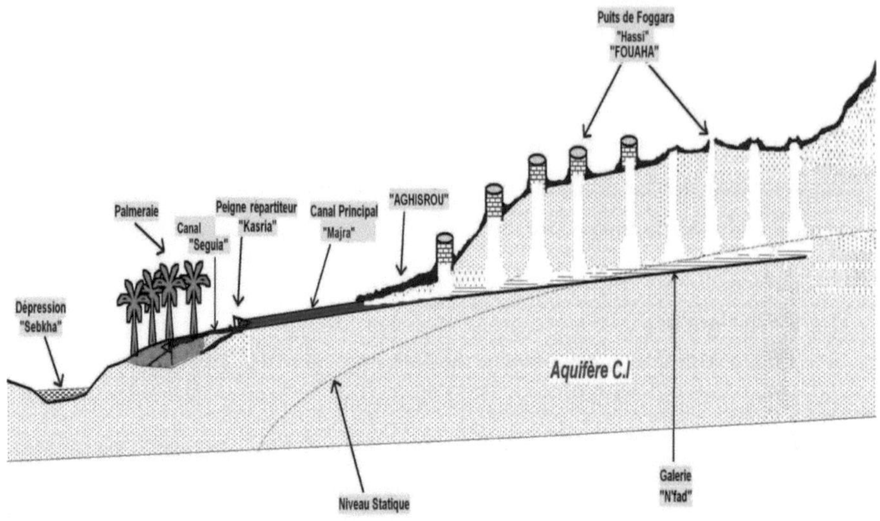

Figure 01 Descriptive diagram of a foggara (Benhamza, 2013)

4.1.2- The "Hassi" well

 The wells are the visible part of the foggara (Photo 02), inside the towns they are an aesthetic object, they are used for access and cleaning of the foggara (removal of excavated material or the addition of fill) and for aeration of the foggara, the distance between the wells is not constant, it varies from one region to another (Gourara, Touat and Tidikelt) and according to the type of terrain crossed, it varies between 7 and 40 m in length. The depth of the wells varies according to the static level of the water table and the topography of the region, shallower in Adrar than in other towns; Zaouiet Kounta and Reggane where the static level is quite deep, the depth of the wells varies between 2 to 40 m, with a diameter of 0.5 to 2 m. The shafts are aligned parallel to the direction of flow of the water table, to increase the flow of the foggara, shafts are added to the left and right of the main alignment of shafts for the second gallery.

The wells are usually plugged to prevent the penetration of sand into the foggara outside the city and the waste inside the towns.

Photo 02 Succession of wells indicating the presence of a foggara

4.2- Distribution

The water is distributed right at the exit of the gallery and is based on the following elements:

4.2.1- Aghisrou

This is the part where the gallery comes out on the surface, it is located between the first shaft and the main channel "Majra", usually covered by cement or rock slabs with clay. The length of the Aghisrou depends on the depth of the first shaft and the ground level (from a few meters to hundreds of meters).

4.2.2- Main channel "Majra".

The main channel "Majra" is a rectangular channel used to conduct water to the "Ksaria" comb distributor, its length is from a few metres to a few kilometres, it is built with clay, currently these channels are made of cement and even PVC, to minimise losses by infiltration.

4.2.3- "Kasria" distributor comb

At the end of the main canal "Majra", the water is distributed by a comb called "Kasria", made of flat stone. The kasria has a triangular basin to store the water before it is distributed among the co-owners. The basin is equipped with a tranquilizer which dampens and calms the flow. The comb has several openings of varying sizes. Several types of Kasria can be seen in the palm grove: the Kasria Lakbira or main comb, the secondary comb and the small multiple comb.

The "Kasria Lakbira" is a triangular shaped basin cut at its base by a tundish acting as a water tranquilizer. It allows not only the damping and calming of the flow before its distribution but also the measurement of the flow rate. The Lakbira kasria receives the entire flow from the Foggara, which is then divided into 3, 4 or even 5 gullies (Seguias). From this triangular basin, the Seguias fan out in all directions and are directed towards the plots to be irrigated.

Photo 03 Kasria Lakbira (main)

Photo 04 Clay Madjen

Photo 05 Seguia

4.2.4- "Madjen" basin

It is a recovery and regulation basin that receives water from multiple kasriates. It is relatively shallow and acts as a water tower. It is located at the highest point of the garden to allow the water to flow by gravity into seguias and irrigate the whole garden. The Madjen is built so that it can be filled in 24 hours. An older earthen Madjen can be seen, the bottom of which is covered with a layer of clay to prevent seepage. The newer Madjen is made of cement. The multiplication and distribution of Madjens in the palm grove creates freshness during the summer thanks to the humidity they release during the day.

4.2.5- "Seguia" gutter

The Seguia, which designates an open canal with a rectangular or circular cross-section, is generally built of earth. An appreciable amount of water is lost through infiltration and evaporation. This encouraged the oasis people to build Seguias in cement to reduce infiltration, but this solution caused the drying out of some palm trees planted near these Seguias. The canals drain the water from the kasria lakbira to the Madjen, then from the Madjen to the Gamoun (garden). As soon as you get closer to the gardens, the Seguias multiply and take different directions, without however overlapping. This tangle of Seguias causes a remarkable freshness throughout the palm grove.

5- Types of foggara

Depending on the geological and hydrogeological context in which they are dug, different types of foggara can be distinguished :

5.1- The foggaras of the Continental Intercalary

This group contains the largest number of Saharan foggaras (Touât, Gourara and Tidikelt). In the sandstone parts of the continental interlayer (Touât in particular), the gallery is narrow, clean and well-hewn and is no more than 0.6 m wide.

In the North of the Touât, the waters of the foggaras cross limestone formations, taking on salts (sulphates, carbonates of lime), which are deposited in layers along the walls of the gallery.

In the southern part and at the Tidikelt, the foggaras are dug into the sandy-clay formation of the intercalary continental and are well cut. The walls crumble, the gallery widens and large caverns are formed as a result of the landslide effect.

5.2- The Foggaras of the Limestone Slab (Tertiary)

On the southern edge of Erg, a number of foggaras are dug into the limestone slab covering the hamada. They are shallow (3.5 m at the upstream shaft) and have a high flow rate.

5.3- The Foggaras of the Quaternary alluvial deposits

The most typical are those of Hoggar in Tamanrasset. They are dug in the arenas and coarse sands of the alluvial deposits of the wadi. The lack of cohesion of the materials, constituting the vault and the walls as well as the damage caused periodically by the floods, gives them a particular ruiniform aspect. The upstream part which is closest to the axis of the wadi is destroyed and filled in, the foggara is then reconstructed about ten metres from the wadi. The succession of wells, of different ages, gives a chaotic and incoherent topographical ensemble which looks very different from the alignments known at Le Touât, Gourara and Tidikelt.

6- Operation of the foggara

The foggara is a draining gallery dug in a straight line from upstream to downstream, which collects and brings underground water to the land to be irrigated, thanks to an appropriate slope. Watering is done by gravity flow and is favoured by the favourable topographical conditions, where the soil level is lower than the piezometric level of the water table of the continental interlayer. The draining or essential part of the foggara is the porous part of the canal or also called draining gallery, it is dug in such a way as to allow the water to circulate and allow the worker to pass through during the construction phase.

The shafts dug along the foggara allow visitors to visit the gallery for maintenance and cleaning. It should be pointed out that the channel situated outside the water table is only used to transport water, unlike the gallery situated inside the water table, which is the useful part of the foggara.

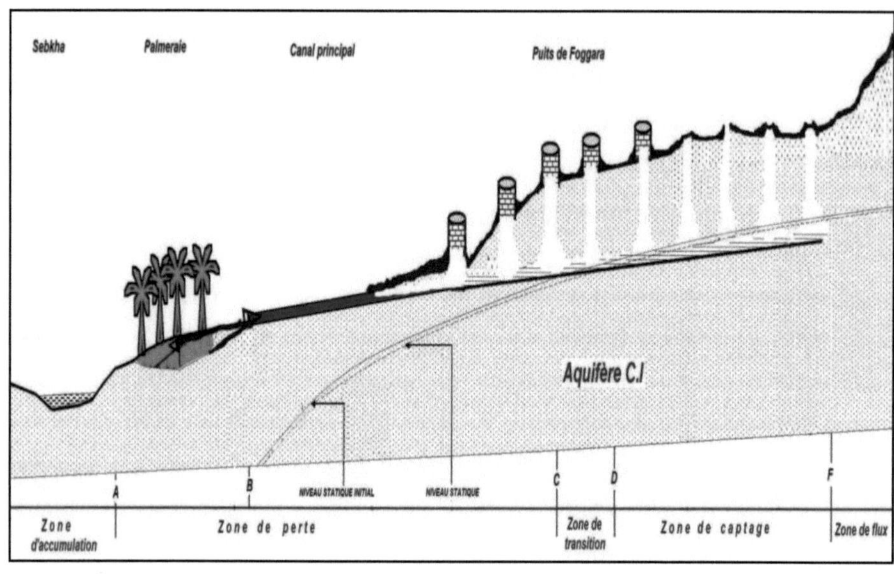

Figure 02 Descriptive diagram of how a foggara works

The static water table level is located above part (D-F) in Fig. 02, the water enters the gallery and moves under the effect of atmospheric pressure and the hydraulic gradient towards part (C-D), stabilisation takes place in this part at the point of intersection of the water table level and the gallery slope, the water from this point flows by gravity towards part (B-C).

In the course of time the water table level drops, the point of intersection moves upstream, from the zone (C-D) to the zone (D-F), the hydraulic load decreases and the flow of the foggara drops.

The foggara has dried up when the flow in point B is zero, i.e. the flow from part (C-F) is equal to the flow lost by infiltration and by

evaporation in the part (B-C).

When the point of intersection of the level of the water table and the gallery reaches point F, the foggara is then "dead".

7- Measurement and sharing of the foggara flow rate

After each cleaning and maintenance operation, a re-measurement and distribution of the foggara's flow is carried out.

Each foggara has a "Zemam" register in which the names and shares of the co-owners are recorded, as well as any modification, purchase, sale or rental of one or more water shares of the foggara.

The Chahed "Witness" who holds the "Zemam" register announces the flow measurement. a main or secondary Kasria at the request of one or more co-owners, or

after each cleaning and maintenance operation of a foggara.

7.1- Measuring tool

The measurement of the flow of the foggara is done by a traditional manufacturing tool called "Louh" in the region of Gourara, "Chegfa or Halafa" in the region of Tidikelt and "Kiel Asfar or Chegfa" in the Touat, it is a copper plate of different shapes and diameters, flat in the Touat of 57 x 18 cm (Photo 06), and cylindrical in the Tidikelt, 15 cm high and 25 to 30 cm in diameter.

The tool is drilled with a row of holes of different diameters that correspond to multiples and sub-multiples of the unit of measurement.

Photo 06 Kiel El Ma with his Louh (Boutadara, 2009)

7.2- Unit of measurement

The unit of measurement of the flow "Habba", "Habba zrig", "Habba maaboud", changes name in each region, it is called kherga in the Tidikelt, Tmen, Majen, Sba, Aud, Kherga in the Gourara and Sbaa, Majen, Habba in the Touat. The unit of measurement is not the same in all regions, it differs from one region to another and even from one foggara to another in the same ksar, hence the different tools used to measure the flow for each foggara.

In the Touat the unit of measurement is about 0.058 l/s, while it is 0.133 l/s in the Tidikelt, table 01 below gives the different flow rates in the Adrar region.

Table IV-01: Foggara flow measurement units according to (Remini, 2008)

Palm grove	Unit	Flow rate l/s
Timimoun	Tmen	0,0261
Deloul	Majen	0,0166
Charouine	Sbaa	0,0833
Tinerkouk	Aud	0,0633
Aougrout	Kherga	0,0683
Ouled Said	Habba	0,0433

The Habba is worth 24 Kirat "Carat "(table 02) and the Kirat is worth 24 Kirat of Kirat (table IV-03), so precious is the water, they used the unit of death of gold "the carat " to share it, the following codification is used for the measurement of the flow.

Table 02 Habba sub-multiples (Benhamza, 2013)

Value	Quantity	Equivalent	Symbol
Habba	1	24 Kirat	-•
Habba Zerig	1	24 Kirat	-•
Habba Maaboud	1	24 Kirat	-••
A Habba Kirat	1/24	1/24 of Habba	•
Two Habba Kirat	1/12	2/24 of Habba	:
Habba Kirat Tois Kirat	1/8	3/24 of Habba	⋮
Four Habba Kirat	1/6	4/24 of Habba	|
Six Kirat de Habba	1/4	6/24 of Habba	|:
Eight Habba Kirat	1/3	8/24 of Habba	||
Twelve Kirat of Habba	1/2	12/24 of Habba	|||
Twenty-four Kirat de Habba	1	24/24 of Habba	||| |||

Table 03 Kirat sub-multiples (Benhamza, 2013)

Value	Quantity	Equivalent	Symbol
Kirat	1	24 Kirat de Kirat	-•
A Kirat	1/24	1/24 of Kirat	-•
Two Kirat	1/12	2/24 of Kirat	--••
Three Kirat	1/8	3/24 of Kirat	---•••
Four Kirat	1/6	4/24 of Kirat	—
Six Kirat	1/4	6/24 of Kirat	— ••

Eight Kirat	1/3	8/24 of Kirat	=
Twelve Kirat	1/2	12/24 of Kirat	≡
Twenty four Kirat de Kirat	1	24/24 of Kirat	≡ ≡

7.3- Flow measurement

The measurement of water depends on the technique of sharing, which becomes the art of measurement.

It is an economy which determines the rules attributing to each entitled person the share which is due to him, according to his contribution, or his right, the measurement of water becomes a symbol of a know-how which will be transmitted from generation to generation.

The "Kiel El Ma ,"the water measurer appointed by the djemaâ of the ksar "village committee in "consideration of his knowledge, honesty and irreproachable conduct, carries out the measuring operation with the help of an accountant "El Hassab "and one or two workers.

Kiel El Ma installs the measuring tool in the Kasria El Kebira or Seghira (the large or small comb) or in a seguia to measure the flow, he fixes the tool with clays and proceeds to close and open the different holes of the tool , each hole corresponds to a fraction of the flow that the meter knows perfectly well, after several measurements he gives the flow of the foggara or seguia to the accountant "El hassab "who makes the calculations of division of the global flow between the shares of the co-owners and counts the share of each co-owner, at this moment Kiel El Ma is doing the work of dividing the flow between the canals, from where each co-owner receives his share of water continuously 24 hours a day, there is another method of sharing the flow called "Nouba) "literally water tower) where each co-owner receives a flow in a determined time (per unit of time), this method is only used in the village of Tamentit for the foggara of Hennou.

Photo 07 "Chegfa "Flow measurement tool (Benziada, 1996)

7.4- Flow sharing

The initial flow of the foggara is given in relation to "Habba Zrig" i.e. the initial flow of the foggara after its exploitation, where each owner has his own flow of Habba Zrig, after several years of exploitation the level of the water table is lowered, the flow drops, then the measured flow of the foggara is called "Habba Maaboud", so it is the real flow of the foggara, while Habba Zrig is the fictitious flow of the foggara.

8- Evolution of foggaras

The foggara is born, water reaches the gardens. Its evolution over time is dictated by the increasing need for water and the expansion of crops. At that time, there was a tendency to multiply the number of wells. But the progression is limited by the terminal cliff of the plateau, and the difficulties increase with the increasing depth of the wells. The well-digger will then try to deepen the drain where the water oozes from the sandstones.

Here too, the difficulty lies in maintaining a sufficient slope to allow water to flow. At the same time, and this is a fact, the depletion of the water table complicates the situation considerably. We then find ourselves in the following alternative:

➤ Deepen the foggara and move the gardens down the depression when possible.

➤ Or abandon the foggara and she dies. In the regions of Touat, Gourara, Tidikelt some foggaras present, when you follow their underground routes, several superimposed galleries.

On the surface, there are remains of the abandoned gardens which correspond to the level irrigation of the various galleries. The palm grove moves and follows the movements of the water.

But there are cases where gardens can no longer migrate into the depression. In such cases, the collecting basins end up below the level of the gardens and irrigation is done by drawing water from the ground.

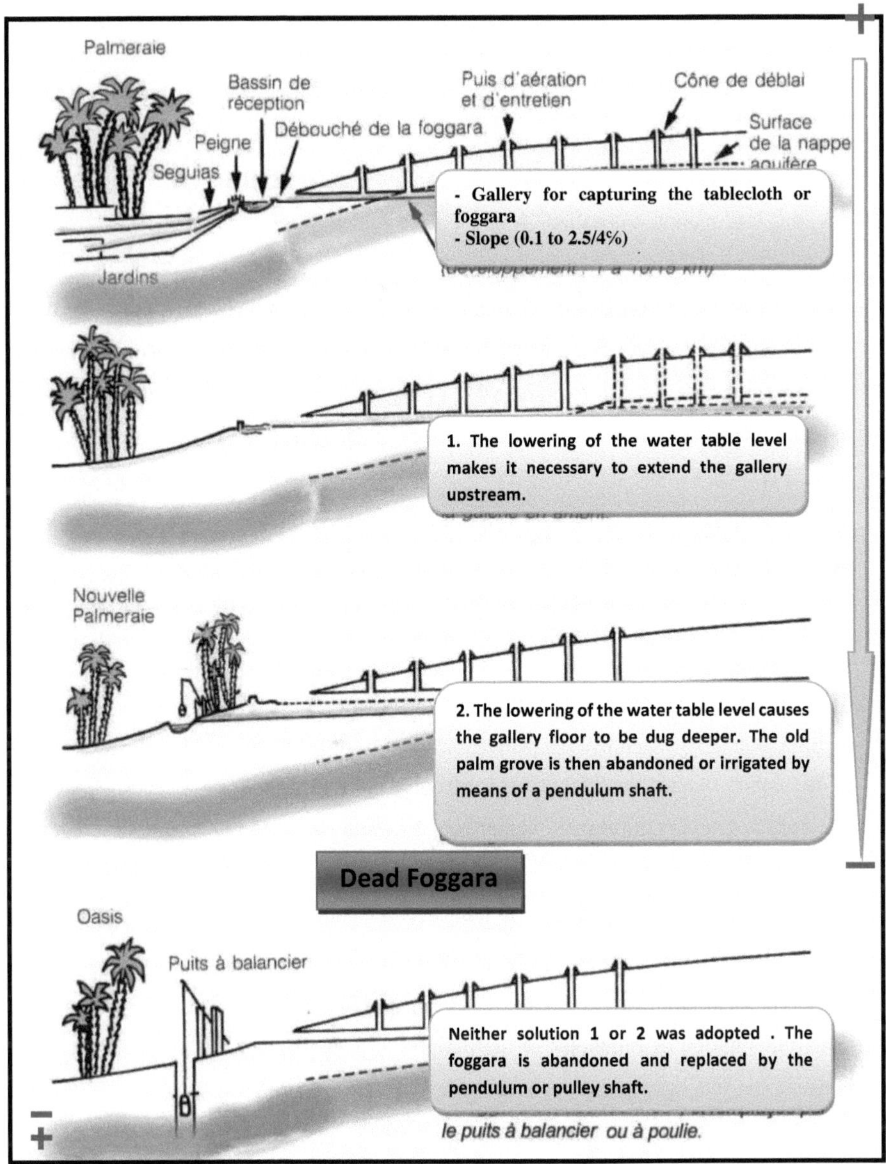

Figure IV-03: Diagram explaining the evolution of the foggara (from Michel Janvois, modified)

9- Cleaning and maintenance of the foggara

The foggara requires periodic cleaning and maintenance of the gallery, this operation is done every year or when an event occurs (flood, landslide), it calls for all co-owners to participate in this operation, the amount of the participation is done according to the debit received, by day work or payment of paid labour. It consists of cleaning the gallery by removing sand and rubble, as well as deepening the gallery in the event of a fall in the flow rate, this last operation is called "Amazer".

To increase the flow of a foggara adds an extension, it is called "Tarha or Ratba", branches are then added to the foggara by adding a "kraâ" (leg), this branch (kraâ) departs from the main foggara to the left or right at an angle of up to 45° and that it must not exceed the limit distance between two foggaras which is equal to 100 Khamas, one (01) khamas equal to 02 m, (Hadj Hamadi and Kobori, 1982). Thus 200 m between two foggaras, one can find several branches in the same foggara.

At present it is rare for a foggara to be extended, given the amount of work to be carried out, i.e. to increase the length of a foggara by 10 m, a shaft of 20 to 30 m must be drilled vertically and a tunnel must be dug horizontally, 10 m long and 02 m in diameter.

10- Foggara degradation factors

If the Foggara has survived for centuries, today it unfortunately finds itself in a situation of decadence, due to the fact that it is confronted with a multitude of constraints, the causes of which are mainly anthropic. Numerous constraints have an unfavourable effect on the functioning of the foggara, and have consequently contributed to the degradation and disappearance of this hydraulic system in certain localities.

10.1- Influence of water drilling on the foggara

The implantation of boreholes in the periphery of the foggaras has caused interference between the two systems through the influence of the more powerful "borehole" capture method on the traditional "foggara" capture method.

The foggara only captures the first saturated metres of the Continental Intercalary tablecloth, any fluctuation in the tablecloth level has a direct influence on the productivity of the foggara.

In this context, Remini points out that the difficulties of coexistence between this traditional system and modern water collection processes are a source of problems and conflicts. Exploitation of the groundwater table by the foggara and exploitation by pumping has led to strong drawdowns since the 1960s (when the first deep borehole was installed in the region) and this generates conflicts.

Indeed, the multiplication of the latter caused the tablecloth to collapse, the piezometric level of the tablecloth falling to remain below the gallery. The draining part of the foggara dries out and ends up becoming inactive.

In addition, the responses of farmers in the region to some of the questions we asked tell us that :

"Foggaras flows are steadily declining, especially in recent years when modern agriculture, based on irrigation from boreholes, is growing considerably, while being encouraged by the state. This state of affairs seriously threatens traditional agriculture and consequently the survival of the foggaras".

10.2- Nature of the exploited aquifer :

The foggaras drain water from the water table of the continental intercalary, the resources of this extensive aquifer system are very low-renewable, i.e. each volume of water extracted influences the overall volume and which then results in the continuous drawdown of the static level of the aquifer. According to the mathematical model SASS (Système Aquifère du Sahara Septentrional) in the wilaya of Adrar, where the water table of the Continental Intercalaire is free over a large area, the piezometric drop observed between 1950 and 2000 is often a few metres: 5 to 20 m in 20 years in the Gourara, 3 to 28 m in 30-35 years in the Touat, and 5 to 10 m in 30-35 years in the Tidikelt. Generally speaking, the tablecloth of the Continental Intercalaire shows a significant drop in Algeria (Ansari, 2004).

10.3- Collapse of the foggaras

It is a phenomenon that occurs through two processes:

➢ A first slow process, during which the flow of water causes a continuous erosion of the gallery bed, which in the course of time causes an enlargement of the section of the gallery, then the collapse of the foggara (Meghier, Sali and Taskhment "Reggane", Boughiol "Zaouïat Kounta"). This case is often repeated throughout the foggaras, especially those crossing towns and roads. Vibrations caused by vehicle traffic and urban development are the main causes of the collapse of foggaras. In general, landslides and rock falls inside the gallery cause a rapid decrease in the flow of the foggara and not a stop in the flow;

➢ A second rapid process, caused by the passage of floods which are generally sudden and of high intensity. The run-off water reaches the gallery through the ventilation shafts, causing the foggara to collapse. For example, in the city of Timimoun, 3 foggaras collapsed following the floods of 2003, while the foggara of Amokrane was abandoned following the collapse of part of its gallery.

10.4- Sanding of foggaras

Contrary to the problem of the collapse which affects the foggaras of the Tadmait plateau, silting poses problems more particularly to the foggaras of the erg which are the most threatened by this phenomenon. The latter find themselves invaded by the sand of the Grand Erg Occidental.

10.5- Pollution of foggaras waters

The water of the foggaras is of good quality, especially that of the foggaras of the erg. In recent years, a deterioration in water quality has been recorded in some foggaras, particularly those of Timimoun.

Foggaras are threatened by different types of pollution. The foggara of Bendraou (Aoulef) is polluted by gas oil coming from a Sonelgaz station. The foggara of Tourfine d'Aoulef is contaminated by waste water from the sceptic pits located near the drainage gallery, while the wells of the foggaras of Adrar and Timimoun have become public dumps.

10.6- Maintenance problem of the foggaras

Since the beginning of the 1960s, this traditional technique has been in marked decline due to the use of modern water catchment techniques. This conflict led to the abandonment of the foggara by the Oasiens. They no longer maintain the old galleries and no longer build new drains. Leaving this hydraulic system to its own devices and faced with aggressive geological and climatological conditions, several foggaras collapsed and others became silted up and deteriorated. It should be noted that the flow rate of a foggara decreases over time and to keep it in a stationary state, it requires continuous maintenance, otherwise it risks drying up and consequently drying up (Photo 08).

Photo 08 Foggara completely dried out (Adrar)

11- Flow improvement and protection of the foggara

Faced with the situation, particularly described above, it is opportune and urgent to provide for an efficient maintenance method, which takes into account local conditions and which can rehabilitate the original flow rates recorded just after the construction of the foggara.

It must be recognised that the initial flow rate can never be reached, especially when the general water table is lowered by continuous operation.

The classic actions to be undertaken and likely to increase the flow of the foggara are generally the following:

➢ To dig the level of the gallery, which would be obliged to lower the level of the land to be irrigated. This choice will undoubtedly substantially increase the flow rate, but for some gardens it is very difficult to achieve this action.

➢ To dig other wells upstream and to extend the draining gallery of the foggara. This choice is only temporary, so as not to stop the irrigation of the gardens downstream, while waiting for a definitive solution to the thorny problem of maintenance and cleaning of the gallery.

➢ By digging other so-called "converging" galleries in the main gallery, this process will make it possible to increase the flow through the water inlets of the other galleries. The increase in flow is proportional to the number of galleries dug.

12- Location of the foggaras of Adrar

The location of the foggara obeys precise topographical and hydrogeological conditions: on the one hand, the water must be present in the subsoil at a depth not exceeding a few dozen metres (for the longest of the foggaras, the upstream wells reach a depth of more than 40 metres); and on the other hand, and above all, the land to be irrigated must be located below the outlet of the foggara. This is the reason why the foggaras oases are aligned at the lower limit of a glacis (Tadmaït plateau), case of the Gourara, the Touat and the Tidikelt, or on a foothill as in the Moroccan Sahara, or on the terrace, or the arenic alluvium, of a wadi with a notable slope, as in the Ahaggar valleys. The following diagram represents the location of the foggaras in the Wilaya of Adrar.

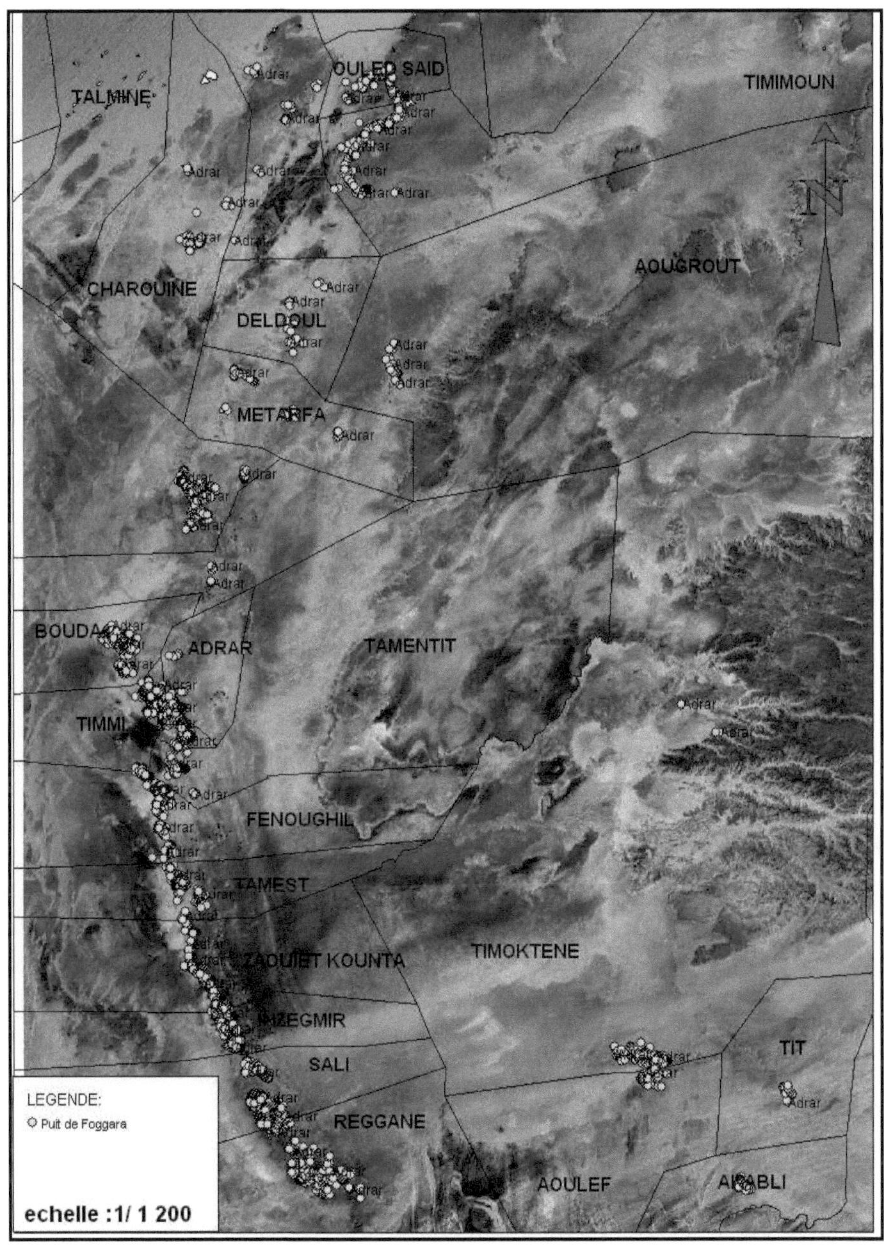

Figure 04 Satellite photo locating the foggaras of the Algerian Sahara (W. Adrar) (ANRH, 2010)

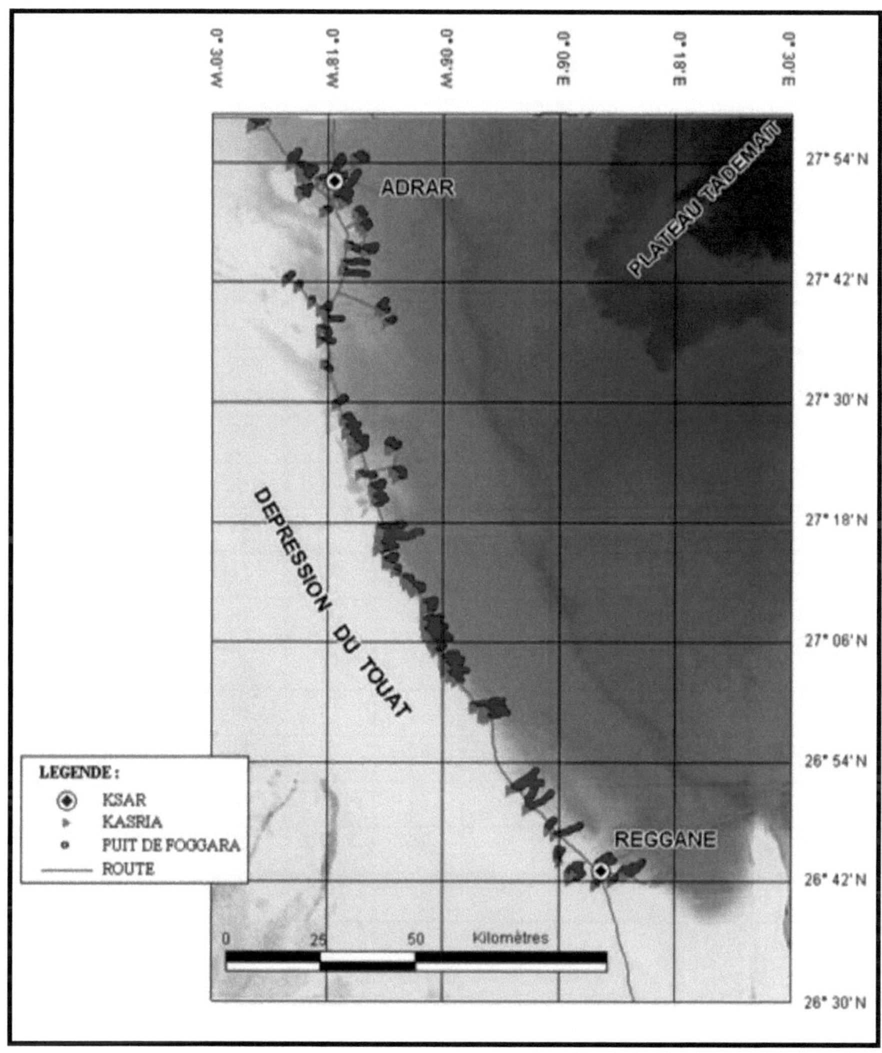

Figure 05 Positioning map of the foggaras in the study zone (Boutadara, 2009)

The analysis of the foggaras positioning map (figure 05), shows the alignment of the foggaras from the town of Adrar to the town of Reggane with a north-west to south-east direction, hydrogeologically this position represents the south-western limit of the outcrops of the Continental Intercalary (C.I.).

13- Census (number and characteristics) of foggaras

Several inventories and flow gauging of foggaras have been carried out, the last inventory in 2011 for the entire wilaya of Adrar.

Table 04 below summarises the inventory of foggaras in the study area (Touat region), which includes 10 communes in the wilaya of Adrar, for the years 1960, 1998 and 2011.

Table 04 Summary of foggaras in the study area (Benhamza, 2013)

Year	1960		1998		2011		Total
State of Fog / Commune	Foggara Perenne	Foggara Tarie	Foggara Perenne	Foggara Tarie	Foggara Perenne	Foggara Tarie	
Adrar	23	0	09	14	05	18	23
Bouda	23	0	20	03	18	05	23
Fenoughil	59	01	40	20	29	31	60
In Zeghmir	64	14	38	40	37	41	78
Reggane	41	07	41	07	36	12	48
Tamentit	50	01	38	13	20	31	51
Sali	31	03	24	10	17	17	34
Tamest	55	10	42	23	23	42	65
Timmimoun	47	05	41	11	29	23	52
Zaouiet Kounta	105	18	96	27	57	66	123
Total	498	59	389	168	271	286	557

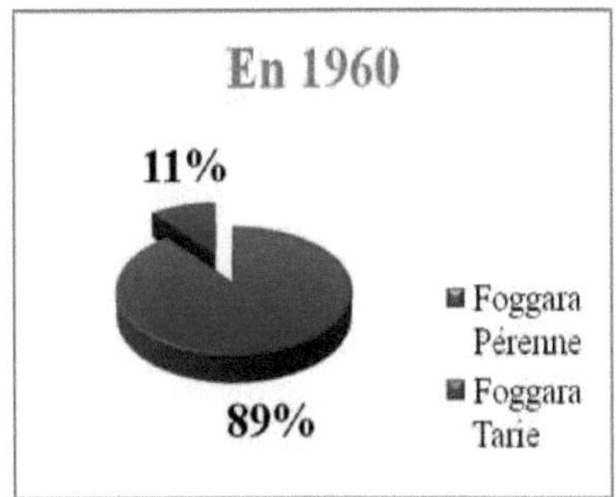

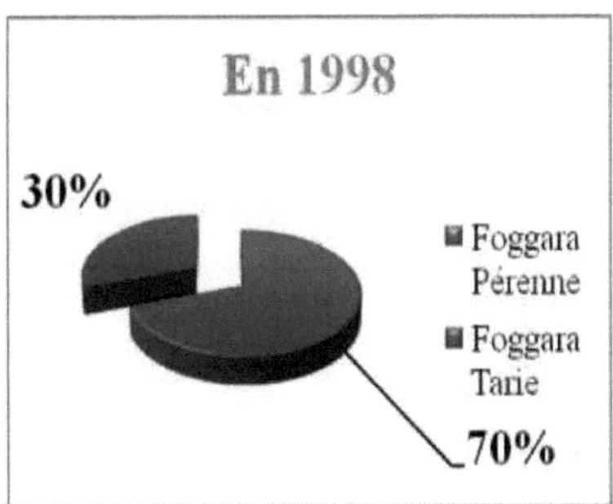

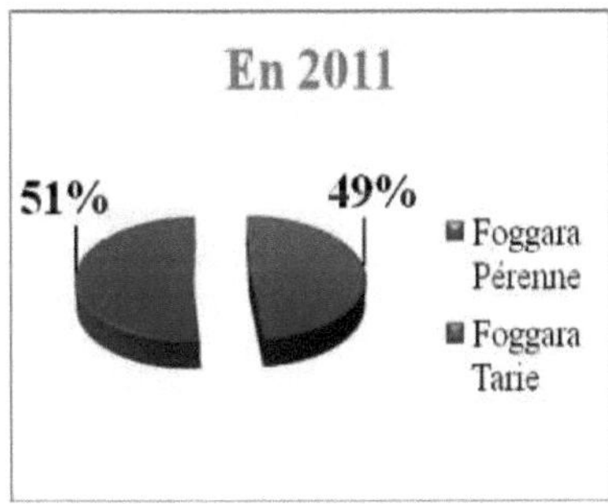

Figure 06 Diagrams showing the state of the foggaras at different dates

In the study region, as in the whole wilaya of Adrar, the general state of the foggaras is unfortunately negative. This is explained by the increase in the number of dry (dead) foggaras from 59 in 1960 to 168 in 1998 and 286 in 2011, compared to the number of perennial (working) foggaras.

13.1- Flow history

The foggaras of Gourara, Touat and Tidikelt are the natural outlet of the Continental Intercalary nappe, according to the results of the last three foggaras flow gauging surveys carried out in the 1960s, 1998 and 2011 (Table IV-05), the overall flow rate of the foggaras has been significantly reduced, it decreases from 3.758 $^{m3/s}$ in 1960 to 1.827 $^{m3/s}$ in 2011, i.e.

a reduction in flow of 1.931 $^{m3/s}$, which represents more than half of the flow rate in 1960 (51.4%).

Table 05 Changes in total foggaras (Benhamza, 2013)

Year	1960	1998	2011
Foggara $^{m3/s})$	3,758	2,846	1,827

For the study area (Table IV-06), the flow rate is 1.909 $^{m3/s}$ in 1960 against 0.918 $^{m3/s}$ in 2011, a reduction of 35.48% compared to the 1998 flow rate and 51.9% compared to the 1960 flow rate.

Table 06 Foggaras flow rate by municipality (Benhamza, 2013)

Year / Commune	1960	1998	2011
	Flow rate in l/s		
Adrar	107,85	75,78	36,9
Bouda	190,9	83,65	70,4
Fenoughil	175,82	159,3	102,19
In Zeghmir	268,17	199,61	178,34
Reggane	148,45	158,38	88,05
Tamentit	180,84	106,6	27,8
Sali	183,92	123,45	88,64
Tamest	155	88,21	86
Timmimoun	212,97	178,63	75,4
Zaouiet Kounta	285,16	249,67	164,54
Total	**1909,08**	**1423.28**	**918.26**

Conclusion

The hydraulic operation of the foggara system is governed by a gravitational flow of water from an area above to an area below from a topographical point of view. The foggara system is made up of two parts, the first is the draining (upstream) part and the second is the downstream part where the water is conveyed in a channel to the distribution network. A specific and well-defined technique for distributing the quantity of water using a measuring instrument: the Chegfa, which makes it possible to allocate to each recipient his or her rightful share of water.

The study of all the data on the foggaras shows that the latter is in a state of progressive deterioration, the drop in flow is confirmed by the taps carried out, with a decrease of more than 50% in the flow and a rate of drying up of the foggaras of 40% compared to the reference year of 1960.

Bibliographical references

ANRH, 2010. National Water Resources Agency, Adrar.

Benhamza M. , 2013. "Hydrogeological and hydrochemical overview of the traditional "Foggara "groundwater catchment system in the Adrar region". Magister memory. Univ. Annaba. 130 p.

Boutadara Y. , 2009. "Étude hydrogéologique des systèmes de captage traditionnels dans les Oasis Sahariennes, Cas des Foggaras de la région du Touat (Adrar)". thesis. university of Oran. 118p.

Hassani I.M. , 1991. "Traditional methods of groundwater catchment in the Algerian Sahara". In : Revue Techniques et Sciences N°6. pp. 20-24.

UNDP, 1986. "Water and the Maghreb: an overview of the present heritage and the future". Report, pp. 131-143.

Remini B. , Achour B. , 2008. "Les foggaras du grand erg occidental Algérien". Larhyss Journal. ISSN 1112-3680, n° 07. June 2008. pp. 21-37.

Printed by Books on Demand GmbH, Norderstedt / Germany